Name: _____  Date: _____

Today is: | Monday | Tuesday | Wednesday |
| Thursday | Friday |

Direction: Read, trace, and write the words:

| 1 | 2 | 3 | 4 |

no. ___  nena

no. ___  formiga

no. ___  avió

no. ___  poma

Name: _____    Date: _____

Today is:  ☐ Monday  ☐ Tuesday  ☐ Wednesday
           ☐ Thursday  ☐ Friday

Direction: Read, trace, and write the words:

| 1 | 2 | 3 | 4 |

no. ___   pilota

no. ___   plàtan

no. ___   bany

no. ___   superior

Name: _____   Date: _____

Today is: Monday   Tuesday   Wednesday
          Thursday   Friday

Direction: Read, trace, and write the words.

| 1 | 2 | 3 | 4 |

no. ___   Hit

no. ___   abella

no. ___   gran

no. ___   ocell

Name: _____ Date: _____

Today is: Monday  Tuesday  Wednesday  Thursday  Friday

Direction: Read, trace, and write the words.

| 1 | 2 | 3 | 4 |

no. ____  manta

no. ____  negre

no. ____  vaixell

no. ____  ampolla

Name: _____

Read, trace, and write the words. Match the images with the traced words by writing the number that is shown on top of each image with the correct word.

### autobús

autobús

Number _____

### nei

nei

Number _____

### marró

marró

Number _____

### bel

bel

Number _____

1  2  3  4

**Name:** _____

Read, trace, and write the words. Match the images with the traced words by writing the number that is shown on top of each image with the correct word.

## atrapar

atrapar

Number _____

## cotxe

cotxe

Number _____

## gat

gat

Number _____

## pastís

pastís

Number _____

1  2  3  4

Name: _____ Date: _____

Today is: Monday  Tuesday  Wednesday  Thursday  Friday

Direction: Read, trace, and write the words.

| 1 | 2 | 3 | 4 |

no. ___ nens

no. ___ rellotge

no. ___ color

no. ___ galeta

**Name :** _____

Trace the words. Cut out the images at the bottom. Match the words with the images by pasting the images inside the boxes.

| vaca | plorar |
|---|---|
| paste | paste |

| pare | dansa |
|---|---|
| paste | paste |

✂ - - - - - - - - - - - - - - - - - - - - - - - - - - - - - - - - - - - - - - - - - -

**Name :** _____

Direction: Match each word to the right picture. Practice writing the word.

gos

cérvols

nina

beure

Spelling Made Easy....

Name: _____  Date: _____

Today is:

| Monday | Tuesday | Wednesday |
| Thursday | Friday |

Direction: Read, trace, and write the words.

| 1 | 2 | 3 | 4 |

no. ___  menjar

no. ___  terra

no. ___  ànec

no. ___  ou

**Name :** _____

Direction: Match each word to the right picture. Practice writing the word.

elefant

peixos

cinc

bandera

Spelling Made Easy....

Name : _____

Read, trace, and write the words. Match the images with the traced words by writing the number that is shown on top of each image with the correct word.

## flor

*flor*

Number _____

## menjar

*menjar*

Number _____

## quatre

*quatre*

Number _____

## granota

*granota*

Number _____

1  2  3  4

# Name : _____

Read, trace, and write the words. Match the images with the traced words by writing the number that is shown on top of each image with the correct word.

## diversió

diversió

Number _____

## nena

nena

Number _____

## anar

anar

Number _____

## cabra

cabra

Number _____

1
2
3
4

**Name:** _____

Read, trace, and write the words. Match the images with the traced words by writing the number that is shown on top of each image with the correct word.

## verd

verd

Number _____

## pernil

pernil

Number _____

## barret

barret

Number _____

## cor

cor

Number _____

1  2  3  4

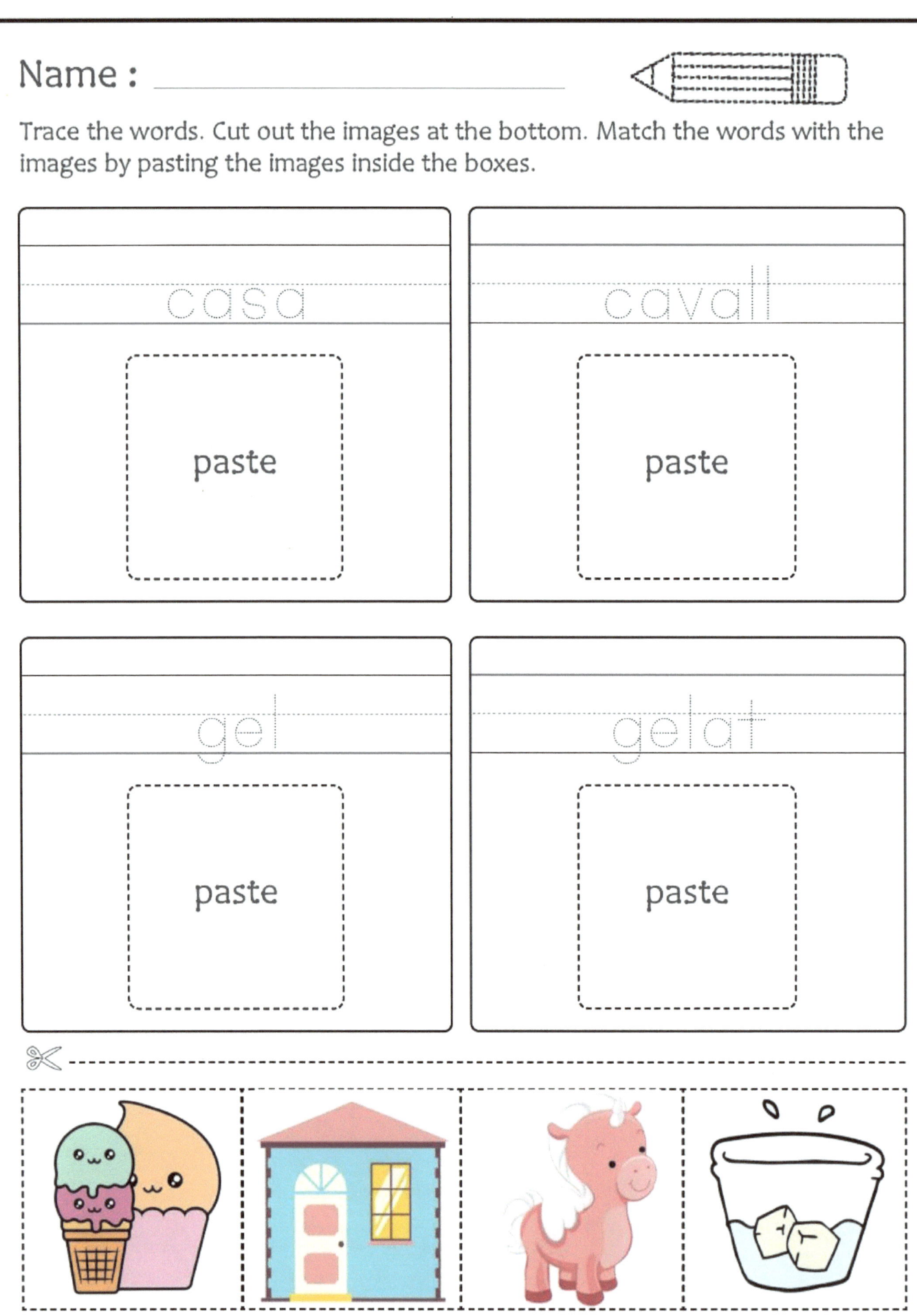

**Name :** _____

Direction: Match each word to the right picture. Practice writing the word.

melmelada

gerro

suc

noi

Spelling Made Easy....

Name : _____

Direction: Match each word to the right picture. Practice writing the word.

peto

lleó

mira

amor

Spelling Made Easy....

# Name : _____

Trace the words. Cut out the images at the bottom. Match the words with the images by pasting the images inside the boxes.

| mare | llet |
|---|---|
| paste | paste |

| diners | mico |
|---|---|
| paste | paste |

✂ - - - - - - - - - - - - - - - - - - - - - - - - - - - - - - - - - - - - - - - - - -

Name: _____  Date: _____

Today is: Monday  Tuesday  Wednesday  Thursday  Friday

Direction: Read, trace, and write the words.

| 1 | 2 | 3 | 4 |

no. ____  rata

no. ____  luna

no. ____  niu

no. ____  nou

Name : _____

Read, trace, and write the words. Match the images with the traced words by writing the number that is shown on top of each image with the correct word.

| nas | un |
|---|---|
| nas | un |
| Number _____ | Number _____ |

| taronja | òliba |
|---|---|
| taronja | òliba |
| Number _____ | Number _____ |

1  2  3  4

Name : _____

Read, trace, and write the words. Match the images with the traced words by writing the number that is shown on top of each image with the correct word.

### pintura

pintura

Number _____

### pingüi

pingüi

Number _____

### pere

pere

Number _____

### rosa

rosa

Number _____

| 1 | 2 | 3 | 4 |

Name: _____  Date: _____

Today is:  Monday  Tuesday  Wednesday
           Thursday  Friday

Direction: Read, trace, and write the words.

| 1 | 2 | 3 | 4 |

no. ___  tirar

no. ___  pot

no. ___  plantes

no. ___  panpra

Name: _____  Date: _____

Today is: Monday  Tuesday  Wednesday  Thursday  Friday

Direction: Read, trace, and write the words.

| 1 | 2 | 3 | 4 |

no. ___  reina

no. ___  pregunta

no. ___  conill

no. ___  pluja

# Name : _____

Direction: Match each word to the right picture. Practice writing the word.

veure

córrer

trist

vermell

Spelling Made Easy....

**Name :** _____

Direction: Match each word to the right picture. Practice writing the word.

seure

ovelles

sabates

set

Spelling Made Easy....

Name: _____  Date: _____

Today is: ⬚ Monday  ⬚ Tuesday  ⬚ Wednesday
⬚ Thursday  ⬚ Friday

Direction: Read, trace, and write the words.

| 1 | 2 | 3 | 4 |

no. ___  dormir

no. ___  sis

no. ___  mitjó

no. ___  cullera

Name : _____

Read, trace, and write the words. Match the images with the traced words by writing the number that is shown on top of each image with the correct word.

**sol**

sol

Number _____

**estrella**

estrella

Number _____

**deu**

deu

Number _____

**tres**

tres

Number _____

1 | 2 | 3 | 4

**Name :** _____

Direction: Match each word to the right picture. Practice writing the word.

tigre

joguina

tren

tina

Spelling Made Easy....

Name : _____

Trace the words. Cut out the images at the bottom. Match the words with the images by pasting the images inside the boxes.

| arbre | tortuga |
|---|---|
| paste | paste |

| dos | paraigua |
|---|---|
| paste | paste |

Name: _____  Date: _____

Today is: Monday  Tuesday  Wednesday  Thursday  Friday

Direction: Read, trace, and write the words.

| 1 | 2 | 3 | 4 |

no. ___  sofa

no. ___  furgoneta

no. ___  gerro

no. ___  vegetal

Name: _____ Date: _____

Today is: Monday  Tuesday  Wednesday  Thursday  Friday

Direction: Read, trace, and write the words.

| 1 | 2 | 3 | 4 |

no. ____  volcà

no. ____  aigua

no. ____  blanc

no. ____  finestra

Name : _____

Read, trace, and write the words. Match the images with the traced words by writing the number that is shown on top of each image with the correct word.

| filats | groc |
|---|---|
| *filats* | *groc* |
| Number _____ | Number _____ |

| zebra | zero |
|---|---|
| *zebra* | *zero* |
| Number _____ | Number _____ |

1  2  3  4

Name: _____  Date: _____

Today is: | Monday | Tuesday | Wednesday |
| Thursday | Friday |

Direction: Read, trace, and write the words:

| 1 | 2 | 3 | 4 |

no. ___  formiga

no. ___  poma

no. ___  avió

no. ___  alligator

Name: _____  Date: _____

Today is: Monday  Tuesday  Wednesday
          Thursday  Friday

Direction: Read, trace, and write the words.

| 1 | 2 | 3 | 4 |

no. ___  pilota

no. ___  llibre

no. ___  timbre

no. ___  nena

Name : _____

Read, trace, and write the words. Match the images with the traced words by writing the number that is shown on top of each image with the correct word.

### gat

gat

Number _____

### blat de moro

blat de moro

Number _____

### vaca

vaca

Number _____

### pastís

pastís

Number _____

| 1 | 2 | 3 | 4 |

Name : _____

Read, trace, and write the words. Match the images with the traced words by writing the number that is shown on top of each image with the correct word.

**gos**

gos

Number _____

**nina**

nina

Number _____

**plat**

plat

Number _____

**cérvols**

cérvols

Number _____

1  2  3  4

Name: _____  Date: _____

Today is: | Monday | Tuesday | Wednesday |
| Thursday | Friday |

Direction: Read, trace, and write the words.

| 1 | 2 | 3 | 4 |

no. ___   au

no. ___   elefant

no. ___   vuit

no. ___   terra

Name: _____  Date: _____

Today is: | Monday | Tuesday | Wednesday |
          | Thursday | Friday |

Direction: Read, trace, and write the words.

| 1 | 2 | 3 | 4 |

no. ___
peixos
_____

no. ___
granota
_____

no. ___
bandera
_____

no. ___
un dofin
_____

Name : _____

Direction: Match each word to the right picture. Practice writing the word.

cabra

pistola

regals

raim

Spelling Made Easy....

Name: _____ Date: _____

Today is: Monday  Tuesday  Wednesday  Thursday  Friday

Direction: Read, trace, and write the words.

| 1 | 2 | 3 | 4 |

no. ___  barret

no. ___  casa

no. ___  pernil

no. ___  cor

www.ingramcontent.com/pod-product-compliance
Lightning Source LLC
Chambersburg PA
CBHW051932210526
45473CB00006B/2225